3rd Enterprise and Infrastructure Resilience Workshop 2021

Online
27-29 September 2021

ISBN: 978-1-7138-4457-0

Printed from e-media with permission by:

Curran Associates, Inc.
57 Morehouse Lane
Red Hook, NY 12571

Some format issues inherent in the e-media version may also appear in this print version.

Printed with permission by Curran Associates, Inc. (2022)

For permission requests, please contact AIChE
at the address below.

AIChE
120 Wall Street, FL 23
New York, NY 10005-4020

Phone: (800) 242-4363
Fax: (203) 775-5177

www.aiche.org

Additional copies of this publication are available from:

Curran Associates, Inc.
57 Morehouse Lane
Red Hook, NY 12571 USA
Phone: 845-758-0400
Fax: 845-758-2633
Email: curran@proceedings.com
Web: www.proceedings.com

TABLE OF CONTENTS

RESILIENT SYSTEMS MODELING

POSTER SESSION

Resilience Requires All Disciplines: The Planetary Limits Academic Network (PLAN)

Authors:
McCall, B. - Presenter, University of Dayton

Humanity's grand challenges – energy and resource depletion, climate change, ecosystem degradation, economic growth and stability, and equity – aren't independent problems we can solve, but are aspects of an unprecedented *predicament*. Our current way of life exceeds planetary limits and cannot be sustained.

Achieving true resilience will require engaging scholars in every academic discipline to develop an integrative understanding of planetary limits, and appropriate responses. This might involve envisioning and laying the groundwork for scenarios that lie between techno-utopian and apocalyptic.

The Planetary Limits Academic Network is a just-launched transdisciplinary community for scholars who appreciate the interlocking nature of humanity's grand challenges and recognize that modernity is incompatible with planetary limits.

If these principles resonate with you, please join us at planetarylimits.net!

1. Humans are a part of nature, not apart from nature.
2. Non-renewable materials cannot be harvested indefinitely on a finite planet.
3. The ability of Earth's ecosystems to assimilate pollution without consequences is finite.
4. Energy throughput is essential to all human activities, including the economy.
5. Technology is a tool for deploying, not creating energy.
6. Fossil fuel combustion is the primary cause of ongoing global climate change.
7. Exponential growth, whether of physical or economic form, must eventually cease.
8. Today's choices can simultaneously create problems for and deprive resources from future generations.
9. Human behavior is consciously and unconsciously shaped by mental models of culture that, while mutable, impose barriers to change.
10. Apparent success for a few generations during a massive draw-down of finite resources says little about chances for long-term success.

A Decision-Making Framework Integrating Fluid and Solid Systems to Assess Resilience of Coastal Communities Experiencing Extreme Storm Events

Authors:
Fahad, M. G. R. - Presenter
Nazari, R., University of Alabama at Birmingham

The precise assessment of flood damage and structural resiliency is of the utmost importance for coastal communities, mitigating risk from repeated extreme storm events. However, most flood resiliency studies have been criticized for lack of accuracy and failing to depict the relationships among the hydrodynamics, structural characteristics, and community preparedness. This work intends to present an inclusive approach for quantifying community-scale flood damage and structural resiliency. Large-scale coastal flooding has been simulated and validated with semi-coupled storm surge and 2D-D inundation models. The depth and momentum components of flood flow were integrated with a newly developed multidimensional flood-damage assessment model, which includes the traditional depth–damage relationship as well as building height, age, configuration, and construction material to calculate the resiliency of structures as a function of recovery time, community preparedness, and level of flood-induced damage. The study region experienced flood depths ranging from 4.91 m to 8.06 m from various hurricane categories with 28.69%, 45.62%, and 92.13% damage to properties. The flood damage and resiliency results were presented via geospatial analytics at property level (i.e., microscale) and aggregated census block group (i.e., macroscale) levels, indicating the developed framework's universal applicability and scalability.

Integrating Uncertainty Quantification in Reliability, Availability, and Maintainability (RAM) Analysis in the Conceptual Stage of Chemical Process Design

Authors:
Al-Douri, A. - Presenter
Kazantzi, V., University of Applied Sciences (TEI) of Thessaly
Currie-Gregg, N., Texas A&M University
El-Halwagi, M., Texas A&M University

Traditional analysis of a proposed process design uses average input values in the performance assessment model, thereby generating single-point estimates. The resulting estimates ignore reliability, availability, and maintainability (RAM) considerations, or assume a fixed value based on prior experience. As a result, a probabilistic view of the impact of equipment unavailability on process profitability is not considered. Recent works have proposed a financial framework for incorporating safety and sustainability considerations in the analysis of proposed designs. Based on this research, we propose a framework to integrate RAM aspects during the conceptual design stage in a probabilistic manner using Monte Carlo simulation. Subsequently, full distribution profiles of key process performance indicators are generated, including system and section availability, annual net profit, and return on investment (ROI). Probabilistic characterization of equipment availability also facilitates the prediction of potential safety and sustainability issues, as more frequent process upsets may result in increased flaring and other potential negative consequences. A modified availability metric, using restoration instead of repair times, is used in this work to incorporate recovery, a key aspect of system resilience. A propane dehydrogenation (PDH) process system is used to demonstrate the application and benefits of the framework. The proposed approach allows designers and decision-makers to comprehensively assess the impacts of equipment RAM characteristics on process availability and economic performance.

Resilience of Processing Systems: Analysis and Synthesis

Authors:
Friedler, F. - Presenter, Pazmany Peter Catholic University

Several aspects are considered in designing processing systems, these aspects are usually evaluated via assessment of expected events on the basis of cause-effect relations, e.g., the determination of the output of an operating unit by a simulation program under known values of its inputs and parameters. However, such evaluation is not appropriate for determining the resilience, since resilience is based on unexpected events in addition to the expected ones. Consequently, the cause part of the cause-effect relation is not known or not effective. Natural way of circumventing this issue is to enumerate all possible "effects" as the set of possible failures.

In the current work, the general formulation for determining the resilience of a system is embedded into the P-graph based mathematical modelling framework. The enumeration of all possible failures is performed by the P-graph based synthesis algorithm, i.e., resilience analysis is done by a synthesis tool. The proposed algorithm is general, there is no limitation on the structure of the processing system (e.g., on the number of loops, raw materials, and products) and on the types of operating units.

Embedding this method into the synthesis algorithm of the P-graph framework, resilient processing systems can directly be synthesized. Since most of the critical infrastructures operate as processing systems, the proposed method can contribute to their analysis and design. Three case studies illustrate the method: a process with a single product, a processing system with multiple products, and a network of industrial plants.

Shifting Baselines May Undermine Shoreline Management Efforts in the United States

Authors:
Correll-Brown, R. - Presenter
Gittman, R., Department of Biology and Coastal Studies Institute
Wellman, E., East Carolina University
Eulie, D., University of North Carolina Wilmington
Scyphers, S., Northeastern University
Smith, C., Duke University
Polk, M., University of North Carolina Wilmington

Rising sea levels, extreme weather events, and unsustainable coastal zone development pose threats to coastal communities. Shoreline development and hardening in at-risk areas can damage nearshore ecosystems, exacerbating existing risks to coastal populations. A comprehensive understanding of changes in response to development, storm events, and sea-level rise is needed to effectively mitigate coastal hazards and promote resilient coastlines. Informed mitigation and adaptation decisions require accurate and current coastal data. Lacking such data, coastal communities risk accepting increasingly degraded coastal zones and making poor management decisions based on shifted baselines. To determine whether human modification of shorelines can be accurately quantified over time, we evaluated shoreline classification efforts in the United States. Using these data, we estimated the current extent of US shoreline modification. However, we found that quantifying shoreline modification over time nationally is currently infeasible due to shoreline resolution changes associated with advances in mapping methodologies and a lack of regularly updated shoreline maps. We coupled this analysis with a survey of coastal planners and managers involved with US state shoreline mapping programs, which revealed that 20 US coastal states have undertaken shoreline mapping projects. However, of the 36 projects identified, only 18 had planned updates. Thus, we recommend increasing the scale and funding for several ongoing shoreline mapping efforts focused standardizing mapping techniques and establishing accurate shoreline condition baselines. Without accurate baselines and consistent updates to data, shorelines cannot be managed in a way that effectively mitigates coastal hazards while promoting socio-ecological resilience in a changing climate.

Addressing Challenges of Industrial Biosolids Reuse in Qatar to Enhance Agriculture Industry

Authors:
Hemachandra, N. - Presenter, Texas A&M University At Qatar
Al-Mohannadi, D., Texas A&M University at Qatar
Sengupta, D., Texas A&M Engineering Experiment Station
El-Halwagi, M., Texas A&M University

Qatar is an extreme arid region with poor soil which makes agriculture activity impossible without heavy energy footprint to produce water and artificial fertilizers. At the same time, Qatar has abundance of natural gas that enables large petrochemical activity. Integration of industrial waste and by-products, such as biosolids, could help reduce emissions and enhance the agriculture sector's resilience by providing an alternative feedstock. Biosolids from industrial cities are currently disposed in landfills due to strict environmental regulations. This work addresses the regulatory limitations that are required to be put in place for the change in the application of the biosolids from land disposal to agricultural purposes. The agricultural focus remains within cash crops that grow in arid and semi-arid regions and the benefits of the usage of biosolids together with the problems associated with its usage are discussed. The paper includes a case study analysis of the composition of biosolids from Shell, Qatar in comparison to the guidelines and limitations from Environmental Protection Agency (EPA) and other countries. Variant treatment methods of bio sludge in general and specific treatment methods of heavy metals if they fail to comply with the regulatory limitations are discussed. The analysis is carried though an optimization model which takes the total cost of treatment, use and industrial crops growth and revenue from biosolids by-products.

Pseudo-Component Approach to Distillation Design for Desulfurization of Biodiesel Produced from Brown Grease

Authors:
Liu, C. - Presenter
Parnas, R., University of Connecticut

Greenhouse gas emissions have become a critical problem. A factor in environmental degradation is coupling world population growth and waste management, including wastewater management. An example of the coupling between energy, population, waste and the environment is FOG (fats, oils and grease), a mixture of fat, wastewater, and suspended solids. This work focuses on converting FOG to biodiesel by extracting the brown grease from FOG and conducting a series of reactions and separations. This operation is a solution for both greenhouse gas emissions and sewer clogs, and also reduces the feedstock cost of biodiesel production. High sulfur content is the biggest challenge of brown grease biodiesel. Sulfur content in fuels is restricted because sulfur is converted to sulfur dioxide, causing significant environmental and health problems.

An industrially feasible process is described to convert brown grease to biodiesel that is green, safe, and effective. The three major steps in the process are esterification, transesterification and vacuum distillation. Several acid catalysts in esterification and base catalysts in transesterification were evaluated. A two-step vacuum distillation was designed to reduce sulfur content to approximately 10 ppm, which exceeds the commercial standard in the USA. A successful demonstration project applied this technique at the Greater New Haven Water Pollution Control Authority (GNHWPCA) and a commercial project in Danbury, CT is in construction. A more effective and energy saving desulfurization process will be developed to complete this investigation. Some ideas and preliminary phenomenon will be presented to introduce advanced desulfurization.

Energy Transition and Decarbonization-Production of Biogas from Biowaste (Food waste) with Wood Chips Towards Low-Carbon Environment.

Authors:
Njoku, O. - Presenter, Rivers State University

Depleting fossil fuel sources accompanied by continuously growing energy demands lead to increased interest in alternative energy sources. Besides, the uncontrolled generation of large amounts of food wastes has resulted in severe environmental issues. This is a major concern in global climate change. As the world drives toward sustainable development in all sectors, there is a need to adopt scientific methods to address the spiraling global environmental concerns associated with the continued generation of wastes especially in the light of limited availability of final disposal sites. Biogas is one of the excellent renewable sources of energy and it is produced from the Anaerobic Digestion (AD) of organic waste: hence, providing a great solution to both energy demand and waste management. AD involves a series of biological processes in which organic matter is broken down and transformed into biogas. Food wastes have a high biomethane production potential owing to their high organic matter contents. The most important parameters on the biomethane production, including feedstock characteristics (nutrient content, particle size, and inhibitory compounds) and process parameters (process configuration, temperature, Ph, retention rate, organic bounding, hydrogen concentration, moisture content, organic loading rate, and inoculums) are fully discussed. Due to problems related to efficient biogas production, an effective methane fermentation of food waste by mixing wood chips with feedstock to minimize the sludge generation during the process is proposed.

About Integration of Renewables in Energy Systems

Authors:
Judbarovski, D. - Presenter

About integration of renewables in energy systems

Illusory belief in Smart Grids really being a Big Folly in a case of large districts' demand and supply, a cost problem to convert such power in competitive artificial fuel because it needs electricity cheaper USD 0.01/kWh, and any other electricity storage is a big cost problem.

It's Time to Talk about Chlorine Dioxide.Studies Show Low Level Chlorine Dioxide Fumigation Mitigates Virucidal and Bacterial Load on Both Surfaces and in the Air.

Authors:
Johnson, E. - Presenter

Sustainability of our education, transportation, food service and business systems can be significantly improved with an easily implementable process of routine decontamination and low level dispersion of ClO_2 in these spaces. The challenge has been the creation of creation method to ensure a specific concentration of aqueous ClO_2 is readily useable and safe. This method exists today and testing in the US has proven its ability to be safe, fast and extremely effective at remediation a space within seconds. Dispensing devices are as simple as an open tumbler filled with 12 ounces of aqueous ClO_2 off-gassing on a desk top and refilled daily. To ultra low volume foggers dispensing wide fine mists of ClO_2 at the beginning and ending of each day.

Patented chemistry is bonded to anionic resin beads that when a chlorite source passes through a bed of this resin the chemical bond of the chlorite source is immediately broken, releasing pH neutral chlorine dioxide.

Our devices allow this oxidation-reduction reaction to occur at the time of demand and keep the reductive and oxidative components separate allowing for stable long term availability & storage.

Test results are available to prove the efficacy of such an approach and is widely used in other countries today. Our process is U.S. patented and invented. Research conducted under grant by the US Department of Defense and follows all 12 Principles of Green Chemistry.

Fabrication of Polymeric Films by Various Processing Techniques

Authors:
Talwar, R. B. - Presenter, Sant Longowal Institute of Engineering and Technology, Longowal

Polymers have some versatile applications which compete well with metals, alloys, ceramic, etc. such as packaging, textiles, household articles, construction materials, etc. Polymers are used in the form of fibers, films, coatings, foams, or molded and fabricated articles and exhibit a wide range of properties in terms of strength, rigidity, toughness, flexibility, elasticity, resilience, optical clarity, chemical and solvent resistivity, etc.

The calendaring process, is the simplest form, is employed to produce continuous films and sheets.

Die casting is a relatively low cost process which consists of converting a liquid prepolymer to a solid object with a desired shape. Acrylics, epoxies, polyesters, phenolics and urethanes are suitable foe die casting.

Film casting is also used to produce polymeric films. In this technique, the solution in an appropriate concentration of polymer in a suitable solvent is allowed to fall at a pre calculated rate on an endless metallic belt so that a continuous sheet of the polymer solution is thus formed on the surface of the metallic belt.

The injection molding process is having the high production rate. In this technique, a definite quantity of molten thermoplastic material is injected under pressure into a relatively cold mold where it solidifies to the shape of the mold. Pressure is applied through a hydraulically driven piston to push the molten material through a cylinder.

Extrusion is one of the cheapest methods available for producing many common plastic products such as films, filaments, tubes, sheets, pipes, rods, hoses and straps, all in continuous lengths.

Energy Efficiency Solutions for a Resilient Future

Authors:
Ulloa, J. - Presenter, Constellation

Energy efficiency improvements are often hindered by budget cuts and finite resources, however investments in these advancements can make impactful long-term effects on reducing energy expenses and improving energy equipment, infrastructure and systems. Constellation is a leading energy provider in the United States that works with facilities to achieve their sustainability goals and develop energy resiliency.

Johan Ulloa serves as Distributed Energy and Energy Efficiency Manager at Constellation and can present case studies highlighting the importance of employing the right energy efficiency solutions in order to reduce expenses and improve infrastructure and management systems. Johan can use his experience in managing large distributed energy projects, commercial transactions and partnerships to speak to and provide guidance around common energy challenges facing facilities and best practices for sustainable energy upgrades and implementation. He can also share his perspective on distributed energy trends he is observing across the industry and his predictions for what is to come.

Modeling a Manufacturing Process of Plastics Films for Packaging

Authors:
Talwar, R. B. - Presenter, Sant Longowal Institute of Engineering and Technology, Longowal

Modeling a polymer process in order to better understand the system, to solve an existing problem or perhaps even improve the manufacturing process itself. Furthermore, a model can be used to optimize a given process or properties of the final product. In order to model or simulate a process we need to derive the equations that govern or represent the physical process. Before we solve the polymer melt process governing equations we must first simplify them by using a set of assumptions. These assumptions can be geometric simplifications, boundary conditions, initial conditions, physical assumptions, such as assuming isothermal or non-isothermal systems, anisotropic materials, as well as models, such as non- Newtonian, linear or nonlinear viscoelastic, shear thinning, or others.

These steps on modeling are:

- Define the scope of the problem and the goals that are to be achieved.
- Sketch the system and define parameters such as geometric parameters (dimensions-diameter, channel depth, channel width and helix angle), processing conditions (such as temperature, pressure, rotational speed of the screw), material parameters (such as viscosity and melting temperature) and boundary conditions.
- Write down the general governing equations that govern the variables in the process, such as mass, energy and momentum balance equations.
- Introduce the constitutive equations that relate the problem's variables.
- State the assumptions and reduce the governing equations using these assumptions
- Scale the variable and governing equations.
- Solve the equation and plot results.

However, the pilot system is too small to be feasible, and must therefore be scaled up for production.

A Concise Description on Plastics- Resilience, Sustainability, and Circular Economy

Authors:
Talwar, R. B. - Presenter, S.L.I.E.T, Longowal, India

Polymers are of great importance due to its high specific strength, modulus, easy fabrication, good resistance to fatigue, non-corrosive ability and economic ability.Thermoplastic composites reinforced through bio-polymers such as starch, wood cellulose, etc. can be obtained.

A sustainable polymer is a plastic material that addresses the needs of consumers without damaging our environment, health and economy. The environmental impacts of plastics are not limited to their end- of- life degradation.Economically sustainable plastics include recycle, reuse, and the degradation of plastics that replace metals, ceramics, etc. with comparable properties.

Economy is an area of the production, distribution and trade, as well as consumption of goods and services by different agents. In general, it is defined 'as s social domain that emphasize the practices, discourses, and material expressions associated with the production, use, and management of resources'.

The circular economy is an alternative to the current unsustainable linear economic model, which involves circularity of resource flow by preventing loss of material out of the system. However, circular economy models go well beyond simple concepts of recycling as it approaches a system level approach that integrates economic activity with environmental and social sustainability. The circular economy concepts have developed from approaches to sustainability and incorporate ideas where products and materials in the system are maintained at their highest value for the longest time possible, removing reliance on non- renewable resources, designing in reduction of waste from the outset, as well as avoidance and elimination of contamination, toxicity and pollution.

Aequor's Treatments Boost Biomass from Renewable Feedstocks (algae, plant material) By up to 40% in Half the Time, Increasing Margins of Biofuels and Coproduct Production.

Authors:
Burzell, C. - Presenter, Aequor Inc

Aequor's unique "green" treatment dropped into algae biomass cultivation systems and plant-based biomass fermenters was validated by DOE and private companies to boost the biomass of common strains of algae and yeast by up to 40% in half the time without the need for antimicrobials or equipment cleaners. This reduces the downtime and operational costs (labor, energy, cleaners, need for genetically modified algae and yeast strains, mechanical equipment upgrades, etc.) and extends the duration of production cycles.

Aequor is currently a member of the DOE's DISCOVR Consortium demonstrating methods to improve algae biomass yields in open pond algae cultivation. Algae remains the feedstock that captures more CO_2 than any other land-based alternative.

Significantly in the yeast fermentation process, Aequor's treatment eliminates the need for antibiotics -- which are widely used and end up in residues sold as animal fee, contributing to the emergence of antimicrobial-resistant bacteria -- the Superbugs.

Aequor's innovations are contributing to the profitability of the biofuels industries, which today total $120 billion and are projected to reach $750B in 2025. Today, blending bioethanol with gasoline at the pump is the fastest way to lower CO_2 emissions in order to reach UN reduction targets. Bioethanol has been the fastest growing U.S. agricultural export for the past 5 years.

Boosting the profitability of biofuel also increases margins of the growing number of renewable co-products made from the algae biomass and plant residues: food (e.g. the meatless burger), feed, nutraceuticals, plastics, materials, chemicals, pharmaceuticals, etc. -- already a $200B industry.

Chemical Industry Infrastructure Resilience Process.

Authors:
Jumbo, T. - Presenter, University of PORTHARCOURT

Chemical Industries are the largest manufacturers of almost everything that falls into the large-scale range and as such it's continued stay in business, is imminent to the society at large.

Chemical plant resilience can be ensured: by improving Safety measures and making the existing ones stringent, by employing the best technology assets, by ensuring that operators, gets regular routine training and retraining so as to measure up to the rhythm of current realities that keeps production plants in good business. The plant must not be situated in susceptibility to any form of inevitable natural disasters.

Knowing that some accidents that occurs in plants are as a result of Engineering hitches that Process Control devices would have mitigated, devices such as SCADA, PLCs, DCS, Fire Detectors, erudite fire fighting equipments would be on site and up to date too. Trained personnel to operate these devices should be available. Constant and sufficient checks / turnaround maintenance would be carried out on the plant. Competent personnel would man every asset on site (that has a hazard potentiality). Frequent conduction of environmental analysis that would foretell natural disasters like flood, tsunamis, quakes and weather studies. Obedience to government policies and regulations should be taken seriously. Have Professional Administrative staff to always balance company's expenses with profit.

If all the above are adhered with zero tolerance to obloquy, docility and negligence, the plant and it's assets would be resilient at least to a reasonable extent as the above-mentioned measures are practicable and results-oriented.

An Injection Mold Design

Authors:
Talwar, R. B. - Presenter, Sant Longowal Institute of Engineering and Technology, Longowal

A competent mold designer must have thorough knowledge of the principles of the mold making as the design of the various parts of the mold depends on the technique adopted for its manufacture. From the manufacturing viewpoint we classify the mold into two parts (i) the cavity and core and (ii) the remainder of the mold is bolster.

The injection mold is an assembly of parts containing within it an 'impression' into which plastic materials is injected and cooled. It is the impression which gives the molding its form. The material may be directly injected into the impression through the sprue bush or the molds containing several impressions it may pass from the sprue bush hole through a runner and gate system before entering the impression. The ejector system in a mold is to provide some means by which molded part can be positively ejected from the core has (i) the ejector grid, (ii) the ejector plate assembly and (iii) the ejection techniques.

It is necessary to provide a flow-way in the injection mold to connect the nozzle (of the injection machine) to each impression. This flow- way is termed the feed system. Normally the feed system comprises a sprue, runner and gate.

In injection molding, hot material enters the mold, where it cools rapidly to a temperature at which it solidifies sufficiently to retain the shape of the impression. The temperature of the mold is therefore important as it governs a portion of the overall molding cycle.

Resilience-Based Recovery Scheduling of Transportation Network Considering Connected and Autonomous Vehicles

Authors:
Zou, Q. - Presenter, Carnegie Mellon University
Chen, S., Colorado State University

Enhancing the resilience of transportation networks (TNs) relies on effective post-hazard recovery strategies. However, little attention has been paid to the heterogeneous users' travel behavior in system functionality quantification and the extensive computational burden in metaheuristic solution procedures. A bi-level decision framework is proposed for the resilience-based recovery scheduling of the TN considering the presence of both connected and autonomous vehicles (CAVs) and human-driven vehicles (HDVs). The lower level quantifies the TN's functionality over time considering different travel behavior of CAV and HDV users. The upper level proposes a novel machine-learning-based optimization approach to solve large-scale network restoration problem. This framework can help better quantify the TN's functionality and support effective and efficient recovery scheduling of large-scale TN in different mixed traffic scenarios. A real-world traffic network in Southern California is provided to demonstrate the proposed framework.

A Systematic Resilience Approach in the Assessment of the Water-Energy-Food Nexus Security Regions with Extreme Climate Conditions

Authors:
Nuñez Lopez, J. M. - Presenter, Universidad Michoacana De San Nicolás De Hidalgo
Cansino-Loeza, B., Universidad Michoacana de San Nicolás de Hidalgo
Sánchez-Zarco, X. G., Universidad Michoacana de San Nicolás de Hidalgo
Ponce-Ortega, J. M., Universidad Michoacana de San Nicolás de Hidalgo

At present, satisfying the needs of water, energy and food represents one of the biggest problems worldwide due to its poor distribution and the social and environmental disturbances that affect its productivity, availability, and future security. To face these challenges, the concept of resilience was introduced, which has gained great relevance in the field of process engineering to be able to identify vulnerable areas and support decision-making for the use of alternatives so that functionality of the system is not affected. Our aim with this work is to introduce a systematic methodology to quantify resilience in economic terms, which allows us to identify the most vulnerable scenarios to different disturbances, such as random failures, personnel errors, or natural phenomena, to propose modifications to the system that allow to cope its interruptions, in this way, a better operation and management of resources can be carried out. Moreover, it is possible to know the operation of the nexus, especially in regions whit extreme climatic conditions where is difficult to satisfy the water, energy, and food needs. The applicability of the proposed systematic methodology has shown that it is possible to identify the most susceptible scenarios, as well as the scenarios where the loss of functionality due to a disturbance is less. Therefore, it is possible to take corrective actions to maintain the security of water, energy, and food in future scenarios.

AUTHOR INDEX